Big Oil
and the
IT
Industry

"Excellent, Easy Reading, and Good Humor . . ."
—Jim Pinto, Author of *Automation Unplugged: Pinto's Perspectives, Pointers, & Prognostications* & *Pinto's Points: How to Win in the Automation Business*

BIG OIL
AND THE
IT
INDUSTRY

Two Giants Shake Hands

Dr. Byron K. Wallace

© 2014 by Dr. Byron K. Wallace. All rights reserved.

The views expressed or implied in this work are those of the author.

No part of this publication may be reproduced, stored in a retrieval system, or transmitted in any way by any means—electronic, mechanical, photocopy, recording, or otherwise—without the prior permission of the copyright holder, except as provided by USA copyright law.

Soft Cover:
ISBN 13: 9781494956110
ISBN 10: 149495611X

Library of Congress Catalog Card Number: 2014900970

Dedication

This book is dedicated to my late mother, Josephine Johnson Willis, who was unable to see me achieve this goal and success but was hopefully aware I was destined to achieve it.

To my wife, Sonja, and my children, Byron II and Breylan, who demonstrated superhuman patience. You have been my inspiration.

To my late grandmother, Madie Lee Hilts-Johnson, and my Uncle Willie R. Johnson, without whom I would not have had the strength to stay on target through this journey.

The history of life, as I read it, is a series of stable states, punctuated at rare intervals by major events that occur with great rapidity and help to establish the next stable era. At the end of the twentieth century, we are living through one of these rare intervals in history. An interval characterized by transformation of our material culture by the works of a new technological paradigm organized around information technologies.

Stephen J. Gould
American paleontologist,
evolutionary biologist, and historian of science

Contents

Author's Note. xi

Introduction to the Author. xiii

Chapter 1: The Dawn of a New Era. 1
Chapter 2: Internet Technology 6
Chapter 3: Petroleum and IT—a Model. 14
Chapter 4: The Role of Big Oil 25
Chapter 5: Big Oil's Influence on Information
　　Technology. 33
Chapter 6: Liberalization in the Petroleum Industry. 38
Chapter 7: Oil Production Influence on IT Outside
　　the US 44
Chapter 8: Information Technology Value 49
Chapter 9: Operation Big Boys 53

Author's Note

All facts, figures, and quoted materials have been verified in my dissertation titled "INFORMATION TECHNOLOGY INNOVATION ADOPTION MODELS WITHIN THE PETROCHEMICAL INDUSTRY" and by various other articles, books, and newsletters. For further information, including questions or comments, please contact me at BYKW231@gmail.com.

Introduction to the Author

DR. BYRON KURT Wallace grew up in a humble, small town in Louisiana, where he lived with his mother, a single parent with three gorgeous boys. Most of his time was spent living with his beloved grandmother and grandfather, who demanded that he get an education and become successful in life.

Wallace's educational journey was always filled with challenges. When he was a young boy, certain folks told him that he was not smart enough to be anything in life, but he proved them wrong.

After high school, Wallace attended Grambling State University and graduated with a Bachelor in Science in Electronics Engineering Technology. Six months later, he began working for Chevron.

In 2003, Chevron allowed him to work in Angola, Africa. While there, Wallace decided to pursue his Master's degree, even though he had already worked full time for fourteen years.

In 2007, Wallace obtained an MS in Information Systems Management from the University of Phoenix. After completing his Masters, he decided to enroll in the doctoral program, thinking, *I am already in the zone, so why not keep moving forward?* In February of 2012, Byron Wallace received a Doctorate of Management in Organizational Leadership, with a specialization in Information Systems Technology.

The research Dr. Wallace did in association with his doctorate degree was rigorous and stressful. It required a tremendous amount of time and many sleepless nights, along with facing various challenges in the process of completing his dissertation. His study consisted of analyzing survey data, and as he says, "To analyze survey data, you need a survey to use first, right?"

It took Wallace three and a half months to find the right survey. Then he still had to get permission to use it within his study, which proposed another obstacle—trying to find the right person to grant him approval, which required a number of phone calls and 2.5 months of hoping and praying before approval was granted to use the survey within his research study.

Moving forward and trying to ignore the frustration that began building, he identified forty-two people within Chevron who met the preset criteria to be part of his study. The young Dr. Wallace thought he could finally breathe. Not so fast, Doc! Out of the forty-two people who were asked to be in the study, only ten actually completed the entire survey.

Further, the surveys he did get back did not reflect the results for which he hoped. His study was centered on the concept that since oil and gas companies produce a large amount of product, they would naturally spend a lot of money, particularly on the field of IT, or information technology. However, the outcome

Introduction to the Author

stated that the amount of production had no bearing on the amount of money invested into IT projects.

As we further explore Dr. Wallace's work, we will understand his intentions for producing this study and what he hopes his reading audience will gain from this project.

Chapter 1

THE DAWN OF A NEW ERA

WHEN I FIRST conceived the idea of writing this book, I asked myself questions like, "Does anyone care to hear what I have to say?" The more I thought about it, the more I realized I needed to get this message out. CEOs, CIOs, managers, and IT professionals need to know what happens when two giant industries shake hands and agree to work together. Most organizations, especially global corporations, tend not to fully understand how the wheels turn that make their lives easy and efficient.

This book follows many of the discoveries that occurred in the United States during the 1800s and how they affected the country's growth as a world player. It was a critical time in our history, and many of the most important pieces of technology that we rely on today were first invented in this time period.

One achievement was the Industrial Revolution that began in 1760 and continued until sometime between 1820 and 1840. By the end of the Civil War in 1865, a new beginning had sparked—the age of machines replacing hand labor. This resulted in mass productivity and completely changed American industries, resulting in countless new businesses. Two hundred and twelve years later, many of these same businesses still exist, including Goodyear, Benz, Singer, Coca Cola, and Colt, and they are now mega-mogul conglomerations.

In addition, many new inventions were created during this era. Some of these were the electric light, revolvers, photographic film, the typewriter, Morse code, pasteurization, the first car, sewing machines, and contact lenses.

Advancements in communications were also on the rise with the invention of the telephone by Alexander Graham Bell in 1876. It soon became vital to all business operations and transactions and became the method of social communication among those in the upper society.

The Industrial Revolution also instigated significant turning points in the practice of agriculture throughout the nation and the world. For example:

- In 1850, commercial corn and wheat belts were invented and initiated what has become our current form of industrial agriculture.
- Many farmers were introduced to the use of mechanical corn shellers sometime between 1800–1850.
- Cyrus McCormick developed the reaper, a horse-drawn machine used to harvest wheat.

The invention of the first steel plow by John Deere and the grain elevator by Joseph Dart also greatly benefitted the agricultural landscape.

The Dawn of a New Era

Farming is essential for our survival; no matter what era we are in, we all need food to eat. Thus, these advances did benefit people. But the rise and demands of machinery and the glittering glow of technological uprising come with a heavy price. By 1940, technology and machinery began to shift our system toward giant industrial farm operations. With all the latest technology helping farmers to work more efficiently, somehow many of the old traditions of farming, including the cultural aspect of helping one another as a whole community, were left behind.

> Before the industrial revolution, corn shocks were hauled in good weather to the barn, and then in harsh winter, the young people went from farm to farm in the evenings making a party out of the husking. The person who husked a red ear—and there were many red ears in the days before standardized hybrid corn—got to kiss his or her sweetheart. This was the cultural, even cultured, way of making work pleasant. It was replaced by a farmer husking corn alone in a cold December field, day after day—a misery, one he was driven to when technology made communal work impossible and obsolete, and when traditional social rituals had lost their significance.
> —Gene Logsdon
> Author, cultural and economic critic, and farmer

The community and ritual experiences were lost when traditional agricultural practices were replaced with industrial farming techniques. Our society also lost the inherited knowledge and spirit that comes with human community, and the relationship between the farmer and the land was greatly weakening.

Despite this great loss, the invention of machinery proved to be successful in the mass production of crops, and many new businesses opened up. Investors and banks lent out substantial amounts of money to these businesses, gaining immeasurable profits in return.

Sadly, most of the businesses were centered in big cities. Many people, drawn to the availability of jobs in the cities, left the farming life. In fact, to this day, people seem to prefer big cities because they have hopes of finding better jobs and quality of life there, rather than tilling the soil and allowing Mother Nature to take charge. (Hmmm, I wonder where the "quality of life" really is.)

As the nineteenth century progressed, individual farmers were replaced with more industrial agribusinesses that could shoulder the burden of mass food production. In conjunction with this rise of industrial farming, smaller, family farms became almost non-existent. Today, food production has become a business only big corporations seem to be allowed to participate in. Crops are seen as commodities and traditional farms as we know them have all but disappeared.

In the nineteenth-century world of technology, scientists were busy creating chemicals that could manipulate crop growth and eliminate unwanted vegetation. However, in our modern times, as scientists have discovered more sustainable agricultural practices, they've also realized that our modern agricultural system is unhealthy for both humans and Mother Nature.

Yet another spark ignited further growth in the industrial realm during the early nineteenth century: Oil was discovered in 1859 on a farm in a county of northwestern Pennsylvania. It is known as the Drake Well. (Some may dispute the actual accuracy of the timeframe, suggesting that the discovery of oil goes much further back.)

The Dawn of a New Era

In the 1800s, the railroad system also became widespread, not only distributing goods but also becoming a new way to travel. However, the use of railways gave little hint that a revolution in transportation methods was underway. Railroads and steam propulsion developed separately, and it was not until the one system adopted the technology of the other that railroads began to flourish.

This brief visit into the past makes it much easier to understand the progression of change throughout the twentieth and twenty-first centuries. We all have witnessed the evolving of our surroundings. There have been changes in the ways people dress and interact with each other. We are now forging the differences in race, color, and stature in life. And we have seen the biggest changes in the advancements in technology, which we now know as information technology or IT.

Chapter 2

INTERNET TECHNOLOGY

WHEN I ENGAGE with people at conferences, software launches, or tradeshows, they often ask me, "What kind of work do you do?"

I respond, "I am the domain admin of the process control network for a large oil and gas company." They then get this deer-in-headlights look, like they are really confused about what I just said. I add, "You really don't understand what I just said, do you?"

"No I do not," they often reply.

I explain that I'm a computer geek. I was once a computer nerd, but I have moved up the food chain a little bit.

To some people, mine is nothing but a fancy computer title. (I'd agree if I didn't know any better.) Yet there's a lot of depth and meaning in that fancy name.

In the same way, the name *information technology* may sound fancy and popular and even intimidating to some who have no idea what it is all about. But information technology has an interesting story, which began even before it became recognized by the modern community.

The words *information technology* first appeared in a Harvard Business Review published in 1958. Authors Harold J. Leavitt and Thomas L. Whisler coined the name information technology, or IT, since they couldn't really decide what to name this new innovation that was clearly rising too quickly.

IT is a branch of engineering that deals with the use of computers and telecommunication equipment to store, retrieve, transmit, and even manipulate data. IT also refers to an entire industry. It is an organization of enterprises or businesses that are responsible and accountable for technology used in planning, designing, testing, and distributing the operations of software, computers, and computer-related systems.

Now that we are in the twenty-first century, society continues to rely heavily on the comfort of technology for the easy, fast-paced life created by our technological advancements. (I call it "the microwave generation.") This generation is in such contrast to our ancestors and utterly different from previous societies.

For example, at one time, carriages and horses were the main source of transportation. They gave way to the steam train, which has long been displaced by the car, and now we have hybrid vehicles that can actually talk and park themselves (that's IT at work in the automobile industry).

Do you remember when workers used shovels to dig holes? Nowadays, high-powered tools and machines replace manual labor. We went from telegraph machines to rotary telephones to push-button phones to, thanks to IT, touch-screen cell phones (Bionic, Droid, Blackberry, iPhone, iPod, iPad, iRobot ... What—iRobot? What's with all the intimidating, robotic, fruity names and all the "i" names again?) No doubt you'll agree that IT has far surpassed our wildest imaginations and has become a vital part of our day-to-day activities.

With that said, as a curious citizen, an observer, and a Ph.D. graduate, I've always wondered how IT affects the oil and gas industry as a whole. Because I have a vested interest in the petroleum industry and know that both this industry and IT are important to the survival of our economy, I ask myself, "What drives IT to change?" In my research, I stumbled upon some intriguing information regarding the subject and also came up with my own theory: in the information age, technology's influence on accessibility to information and information processing improvements drives change.

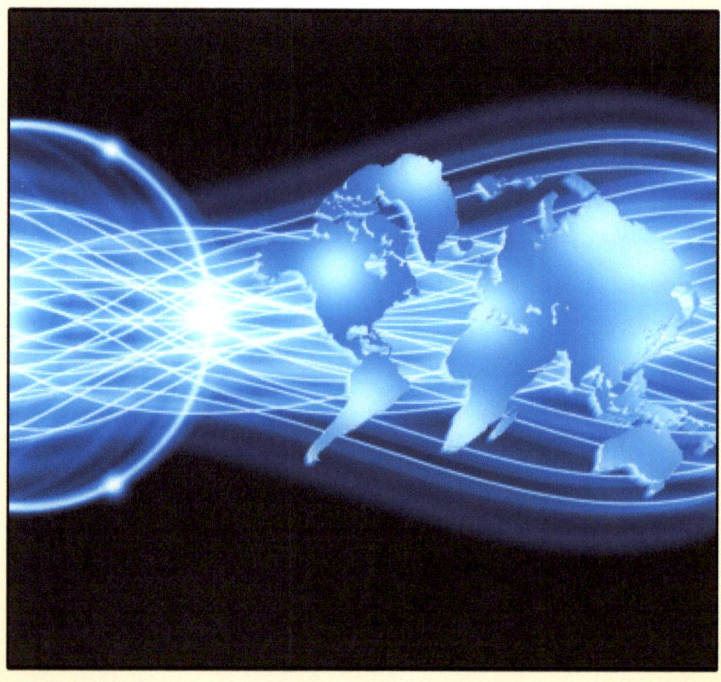

Before we get down to the specific issues, one important characteristic of technology must be pointed out. The rate of technological improvements allows greater automation, which allows workers to work more efficiently and safely. Like many new developments, the use of technology, specifically informa-tion technology, has at times threatened to become a fad. I call it the "Batman-saves-the-day-in-a-collapsing-society-of-Gotham" fad.

Indeed, if you've watched any of the Batman movies, you well know that Gotham was a collapsing, corrupted city infested with crime and poverty. Despite its flaws and challenges, with the help of the few good men in power, Gotham was also on the rise for new developments, starting with the train system that connected the two separate sections of the city (the rich and the poor). One man's vision changed a lot of things in Gotham's society. And we can see the ingenuity of using technology as the Cape Crusader utilizes gadgets and high-tech vehicles to help restore the city. (Of course, we can't forget the villains who tried desperately to cause havoc, using some very sophisticated weaponry of mass destruction.) The rise of technological advancement has become the beacon of the New World that will, no doubt, surpass the greatness of ancient history.

Technology on the Rise

Between 1939 and 1942, Iowa State University Professor John Atanasoff and graduate student Clifford Berry built the world's first electronic-digital computer. What level of intellect could attain the ingenuity necessary to create such an amazing invention? It represented several innovations, including a binary system of arithmetic, parallel processing, regenerative memory, and separation of memory and computing functions. If Professor

Atanasoff were still alive today, I'm sure he would be amazed, even dumbfounded to know what his invention has spawned.

Computers were once just used as problem solvers, but now they are being used as information-processing systems, communication systems, and control systems. The latter uses are called "real-time applications." Central to these processes are a complete time record and maintaining full control over sequence of operations.

Here are some examples of real-time information processing systems: the airline reservation system, communications between airport control towers, and the Stock Exchange computers. Each of these systems can handle a large volume of transactions from diverse points to bring large inventory files up-to-date and process information needed for each transaction.

The faster the computer and its applications are, the better the IT. Previously, only businesses and large corporations were in need of faster, more reliable operating systems and Internet service. However, in today's society, individuals have access to the latest and greatest. Children as young as two years old, teens, adults, and even senior citizens are using iPads and touch-screen computers and cell phones to access the world of cyber space and text people around the globe. Notice how fast our technology has changed, and it will continue to change and get even better.

I have to admit that, like many, I am guilty of being sucked into texting, surfing the net for long periods of time, and checking social networks and emails. Some people even go so far as talking or texting while driving (something I *never* endorse). This is a newly acquired habit by our so-called modern generation that's hazardous enough to take as many lives as drunk driving. It is sad indeed. Although our government and its law enforcers do their best to implement actions that will lessen such tragedies, the death toll continues to climb. As a result, they must also rely heavily on IT to access data and conduct surveys that contribute to careful planning and designing of new ways to make our society safer. This shows that technology is in demand in our everyday lives. This drives IT to change rapidly.

In addition, in the last generation, academia, managers, and professionals have expressed great interest in understanding the contributions of IT innovation within companies. And globally, leaders of corporations are continuing to increase spending on technology. The proportion of total corporate spending allocated to technology has risen tenfold since the mid-1960s. Spending in the USA is growing faster than previously thought, and it is evident that the industry recovery is well on its way.

However, it must be noted that the Obama administration has been trying to hold federal IT spending steady while increasing government efficiency through the use of cloud computing and mobile devices.

> By doing more with less, the administration is driving savings across government and using those savings to reinvest in information technology and services that benefit the American people.
> —Barack Obama
> Forty-fourth President of the United States

IT spending includes purchasing computers, protecting government data, updating websites, and hiring employees who provide technical support. According to *Bloomberg Businessweek*, government IT expenditures rose 7.1 percent a year on average from 2001 to 2009 and have effectively been halted, with no growth from 2009 to 2013 (according to budget). In contrast, the civilian sector IT spending has increased 1.1 percent—to $41.7 billion from 2009 to 2013.

Future Growth

Fueled by an accelerating move to cloud computing and by a boom in associated telecommunications services worldwide, IT spending is increasing somewhat faster than expected. Consider that $3.6 trillion was spent on information technology in 2012 alone. That is a 3 percent increase from 2011. The increase is significant because it happened in the face of a financial crisis in Europe, the slow growth in the United States, and a slowdown in China's economic growth.

By far, the greatest amount of spending in the USA is in the telecommunications industry, with $1.69 trillion spent already in the first half of 2013. Andrew Bartels, an analyst at Giga Research, estimated that tech spending will rise by about 2 percent this year and that investments in the new technology should grow by around 6 percent annually for the next four years.

Steve Milunovich, an analyst at Merill Lynch, thinks that tech industry evolves through waves lasting ten to fifteen years, as innovation leads to a burst of spending on new technologies and then to a less exciting period of assimilation.

Currently, no single model of implementing IT innovation within an organization exists. New IT innovation implementation requires commitment and support from all related levels of an organization. All phases of implementation from initiation to post evaluations within the organization are crucial aspects of its success.

Internet Technology

> For the wit and mind of man, if it work upon matter, which is the contemplation of the creatures of God, worketh according to the stuff and is limited thereby, but if it work upon itself, as the spider worketh his web, then it is endless, and brings forth indeed cobwebs of learning, admirable for the fineness of thread and work but of no substance or profit.
>
> —Francis Bacon
> *The Advancement of Learning*

As I researched and observed the dramatic changes regarding IT, a small, humble grin made way at the corner of my mouth. I realized how fortunate I was to have picked such a career and invested time and great effort to further my studies. Sure, the wages have become somewhat stagnant. However, with the way this world is gearing into high-tech modernization, IT techs have a bright, shiny future ahead. Therefore, I can honestly say, "IT is my friend."

Chapter 3
PETROLEUM AND IT—A MODEL

MOST PEOPLE DO not understand the roles that oil and gas play within the economy. However, as an employee of a petroleum company, I know all too well the effects of oil prices. For us, oil prices are the single biggest factor governing our every move. Things as simple as going to training for new technology depend on oil prices.

On one occasion, I asked my supervisor if I could attend a CISCO certification class for routers and firewalls. His response was, "We don't have any money within the budget this year for training." The underlying story was that since oil prices were low, cuts had to be made, and the training budget had been eliminated. So once again, oil prices had dictated to us how we were going to conduct business.

But oil prices don't just affect petroleum companies. The rise and fall of oil prices has an adverse effect on how organizations outside of the oil industry conduct their businesses too.

Petroleum and IT—a Model

One morning, I sat in my home office caught up in awe of my world. I was completely amazed by its complexity and the great achievements this new era has accomplished. I pondered the fact that those achievements will just increase as time goes on.

It actually didn't help when a thought flashed through my mind while watching the opening ceremonies of the Olympics in London. The presentation was great, full of flashy displays of brilliance that included a fireworks display. Clearly, tones of money, time, energy, and intelligence were put together to create such a striking presentation, and of course, the technology that was used stood out. But what really caught my attention was the adaptation of how life in the early days of our history takes center stage.

At first look, I thought I was seeing *The Hobbit*, with the swirling hillsides covered in patches of green grass. However, rather than seeing a tiny, cozy hobbit home, I saw a rather huge tree that was placed in the center, with people portraying business folks. First, there were men and women working in coal mines. Then farm and factory workers came marching out. I sat in silence, patiently waiting to see what the director was thinking. But it didn't take long before it hit me, and I couldn't help but smile as I thought, *The Industrial Revolution is being depicted.*

I couldn't resist going back and reminiscing about the time when electricity, cars, Wi-Fi, and dish networks were all unheard of. Without my permission, my mind travelled back in time to the era when oil was used to light up torches and lamps, people were less engrossed in material things and less dependent on technology, and life was simpler.

Because I grew up in a small town, I think I know what simple living means. Of course, I didn't always like the idea or understand any of it when I was a young lad. But sure enough, being in certain situations back then made me realize the value of and how to appreciate the things I have now.

As I continued to reminisce, a nagging thought about how all of this came into play frolicked in my head, like an itch that needs to be scratched. And of course, my unusual obsession with the Industrial Revolution came in handy, as it always does. It has continually been a pleasure for me to delve into the fact that it was indeed around that time that the oil industry was born and to consider the role oil production has today.

In the United States, the first oil well drilled was located in a small farm town in Pennsylvania. "Colonel" Edwin Laurentine Drake (1819–1880), a man with no knowledge in geology or engineering, who had faced a life of hell on earth as a parade of disappointments overtook his life, was the man in charge. Prior to this, his collapsing dreams of heroic achievement had forced him to resign himself to a more practical field of work as a farm laborer and night shift steamboat clerk.

While Drake was on a sick leave due to some form of arthritis, George Bissell, a local banker with good connections to people in a higher echelon of society, established a small partnership with companies that extracted oil on a commercial scale in Pennsylvania. Bissell believed that Drake was the man for the job. Despite his gloomy career history, Drake accepted the offer and undertook the leadership. Drake was given a new, dignified title: "Colonel." This was part of Bissell's brilliant way of marketing to attract investors; although, in reality, the only uniform Drake had ever known was that of a railroad conductor.

George Bissell and Jonathan Eveleth soon founded Seneca Oil, originally called Pennsylvania Rock Oil Company. The company was created after the two gentlemen caught wind of reports of petroleum collected from an oil spring in Titusville, Pennsylvania, that was suitable lamp fuel. Before people discovered that oil from the ground could be used, whale oil was the primary source of fuel.

Unfortunately, while big money was being discussed, disagreements and feuds were also relevant. These eventually caused the company and its shareholders to split in half. (Sadly, this still happens today.)

In the spring of 1858, Drake was hired by the Seneca Oil Company to investigate oil deposits. Using a salt well drill and a steam engine to power it, he and the workers began digging in earnest. Drilling holes through layers of gravel wasn't an easy task. When they reached sixteen feet deep, the side of the hole began to collapse, causing great despair to most workers, but not to Drake.

Unfortunately, when the project looked hopeless and money seemed to have been wasted, Seneca Oil Company decided to pack up and leave, abandoning Drake and his men. Drake, however, continued. The man not only invested his own money but also his energy, efforts, and brain, all of which lead to the invention of the "drive pipe." He pursued the drilling against all odds. Truly the man believed in the saying, "If there's a will, there's a way." Of course, Drake had help from his friends, at least the few who believed in him and thought of him as sane rather than insane. Happily, Colonel Drake's hard work and persistence made way for the birth date of the oil industry boom.

Imagine the hype that took place once petroleum hit the market for the first time and how it affected the economy. It must have been nice to be in on the "ground floor."

Big Oil & the IT Industry

I don't think the petroleum boom was any different than the technological boom we're currently in. Petroleum was the "in thing" then. No, let me take that back; petroleum still is a major player today. Although a Technological Revolution is on the rise, the demand for the use of oil and gas remains as strong as ever. Barfield, Raiborn, and Kinney argue that the aggressive alteration of the petroleum industry across the world controls future ability to meet the demand for petroleum markets more efficiently and in a cost-effective, green manner. (I wonder if by saying "green manner" they are talking about Mother Nature.)

The petroleum industry also holds the possibility for greater IT innovation in the transporting, packaging, and delivering of petroleum products, which can generate cleaner, more dependable, and moderately priced energy. Within recent years, the petroleum markets have been liberalized or are being liberalized across the world.

The state of the petroleum industry consists of perpendicularly incorporated monopolies, which have distinct service territories and sheltered consumer bases. The petroleum itself is used for numerous products, in addition to serving as our world's primary fuel source. Due to this, as the price of oil goes up and down, it can cause many to enter a state of panic. Everyone knows that when the price of oil spikes up, everything else does too. (Well, maybe not everything … high prices, low wages—does this ring a bell?)

The processes and systems involved in producing and distributing oil and gas are highly complex, capital-intensive, and require state-of-the-art technology. The importance of the role of IT to continue operations, as well as improve efficiency, continues to grow.

Petroleum and IT—A Model

Various factors contribute to IT innovation within the petroleum industry, such as competition, government control and regulations, national and international pressure, marketing forces, and the need to improve efficiency. The challenges of the petroleum industry across the world are not simply limited to meeting the demands of the market; the monopolistic nature of the industry also plays a large part in the decision-making process.

> Almost everybody today believes that nothing in economic history has ever moved as fast as, or had a greater impact than, the Information Revolution. But the Industrial Revolution moved at least as fast in the same time span, and had probably an equal impact if not a greater one.
> —Peter Drucker
> American (Austrian born) management writer, 2005

Back in 1870, a corporation in Ohio was established. In fact, it was the largest oil refinery in the world and the first multinational corporation—that is, before our supreme court broke it apart in 1911. John D. Rockefeller, American industrialist and philanthropist, was the founder and chairman of the corporation, Standard Oil Company, and also was the richest man in modern history.

Standard Oil Company grew exponentially, gaining an aura of invincibility. Despite its rivals and the chaos in shipping, Rockefeller's dream never ceased. His aggressive business tactics made his competitors overrun by fear of being bankrupt if they refused his offers. It was real hell on earth doing business with him because he was such a merciless, iconic figure, not only during the oil industry of his time but also to the modern industry as a whole. While keeping oil prices low to stave off competitors, the company soon added its own pipeline, tank cars, and home delivery.

Rockefeller was monopolizing oil. Underselling was the company's most potent weapon against its rival companies. Standard Oil's share of world oil refining topped out at about 90 percent. This included pipeline and oil tanks used for both road and railroad transportation. In addition, in an age before the real-time information agents came into our lives, Standard Oil monitored every country, tracking all retail prices and gas sales, and studying the behavior of all the local competitors. It also paid spies to look out for reports of new discoveries regarding oil and how its value could be affected.

In 1911, the Supreme Court ruled that they had found the company guilty of illegal practices. Standard Oil Company was eventually broken up into thirty-four new companies, some of which are well known today: Amoco, BP, Chevron, Exxon, and Mobil.

Before we go further, I'd like to point out that there are two types of drilling methods—one for drilling onshore and one for offshore. Onshore exploration relies on fixed or mobile drillings rigs. Whereas, offshore exploration can require a number of different types of drilling rigs, which include fixed offshore Jack-up drill rigs, Deep Water Drill Ships, and Semi-Submersible drill rigs. These are helpful in geological surveys, which are conducted using various means like testing subsoil for onshore exploration and sophisticated technology such as seismic imaging for offshore exploration.

Petroleum and IT—A Model

Modern science and technology have made oil exploration, which encompasses the processes and methods involved in locating and discovering potential sites for oil drilling and extraction, more productive and highly efficient. Exploration costs can vary dramatically. The cost for unsuccessful exploration, including seismic studies and a dry well, can run $5 million to $20 million per exploration site and, in some cases, much more. However, when an exploration site is successful and oil extraction is productive, exploration costs are recovered and are significantly less in comparison to other production costs.

Since the 1960s, the USA has debated the pros and cons of offshore oil drilling, due to legislative actions throughout history and accidents resulting in oil spills. This debate has become a central political issue, with many factions, both political and social, wanting to place sanctions to ban oil drilling within United States for as long as possible. However, those who justify oil drilling within the United States argue that by producing more oil domestically, America relies less on foreign oil imports and avoids the economic threat from oil trading countries. It sounds pretty appealing, if I may say so. After all, who in the world wants to face such economic turbulence?

Worldwide Spending

The global challenges for the petroleum industry are not simply limited to meeting the demands of the marketplace. They also include the monopolistic nature of the industry. This is one of the greatest hindrances to IT innovation and performance improvement. Various sources indicate that the performance of petrochemical organizations can be improved when there are certain changes in how the government deals with these monopolistic companies. The sources suggest the abolishment of monopolies and the introduction of competition in the marketplace.

Despite the monopolies, economic volatility, oil crises, and decreased employment utilization rate, there is good news. In the midst of a domestic downturn, IT spending is expected to grow by 4.9 percent by the end of 2013, according to a latest report by International Data Corporation (IDC). The stat revealed the IT market is expected to grow to US $2.06 trillion in 2013, up from $43.57 billion in 2012. Also, large government spending in areas such as e-governance will catalyze speedy development of the domestic IT market. And as far as retail and wholesale, energy and utilities, or even healthcare are concerned, they too will experience such growth in IT spending. Corporate businesses are prioritizing IT spending just like investments in sales operations.

However, even though the USA is having a major turnout of good juju when it comes to IT's future, other countries aren't as excited. Worldwide IT spending is forecast to grow, but at the same time, IDC also stated that various regions are facing very different conditions, as various macroeconomic factors have made an impact on the backing industry. For Europe, Compound Annual Growth Rate (CAGR) for the forecast period now stands at 3.6 percent, compared with 4 percent in the previous forecast. IT spending forecasts are essentially unchanged from the previous forecast for the Asia/Pacific region, stated the IDC report, adding that over the forecast period, spending will increase with a CAGR of 7.9 percent. The Middle East remains a strong growth market for financial services technology. And the adoption of public cloud services among US banks has surfaced as a way to innovate, despite tight IT spending controls.

Compared to other areas of the continent, the region of East Africa is quick to adopt technology. Uganda, Kenya, and Tanzania appear to have de-emphasized ICT—and plan to increase budget spending for the next twelve months. Uganda's US $4.8 million ICT sector allocation is the lowest in the past three years, according to an analysis by the Collaboration on International ICT Policy for East and Southern Africa (CIPESA). Finance ministers reviewed the budget blueprints on June 15, 2012, and found that in Uganda, the funding amounts to only 0.13 percent of projected government expenditures over the next twelve months. The country spent $7.1 million in 2011 and $5.7 million the year before that.

Kenyan Finance Minister, Robinson Njeru Githae, did not say a lot in his budget speech as far as the Information Communications Technology or ICT sector goes, but he allocated some $5.6 million for the purchase of computers for schools and removed import duty on computer software.

In Tanzania, which reported far less money in comparison to its neighbors in the past, actually increased duty on mobile telephone airtime, taking it into a league that Uganda has long dominated. There, telephone services are taxed steeply.

Information technology is the backbone of the petrochemical industry. The petrochemical industry faces various challenges during implementations and adoption of IT innovation projects within organizations. These challenges vary among the operating companies within the petrochemical industry. The main challenges faced by petrochemical organizations toward IT innovation adoption and implementation are the lack of support and the resistance to change from management within the corporations. However, petrochemical organizations have matrices of functional groups and teams to ensure adoption of IT innovation within the organization.

Information technology innovation adoption and implementation within the petrochemical industry requires a commitment from all levels of the organization, including executive management. Currently, existing petrochemical companies operating internationally adopt and implement more IT innovation projects than the petrochemical companies operating primarily within the United States.

Chapter 4

THE ROLE OF BIG OIL

EVEN THOUGH DRAKE was the first pioneer of oil exploration in the United States, the modern industry actually began in the region of Mesopotamia, which today probably boasts the largest reserves in the world. The current oil export is more than two million barrels per day (bpd) and Iraq hopes to reach its target of ten million bpd in 2017.

One thing most people do not understand is that there's a difference in the types of oil when it comes to developing budgets. What is referred to as "normal oil" is an oil location that produces around 2,000 or so barrels of oil per day. What is known as "big oil" is an oil location which produces somewhere around 238,000 barrels of oil per day. (These are usually oil locations outside of the US.) When you work within an organization that produces big oil, budgets are pretty much unlimited (especially IT budgets). While I was working in Angola, Africa, which is known for big oil, I personally had a $25,000 signature authority, meaning I could purchase anything up to $25,000 without needing approval from anyone.

The petrochemical industry is part of the biggest businesses in the world: the energy industry, which is estimated to generate 1.72 trillion US dollars per year. The World Energy Council predicts that global energy investment between 1990 and 2020 will add up to $30 trillion, with energy prices reflecting 1992 rates. The growth of demand for petrochemical companies by the automobile industry and various other industries is increasing with the rising population of the globe, innovations, and supply of energy-consuming products.

The petrochemical industry is a complex industry. The process starts with production from fossil fuel, which is a technically and chemically intricate activity. Raw materials in the form of oil fraction natural gases that are obtained from fossil resources are first acquired. The oil and natural gases are then segregated by using highly developed technologies and various refineries, utilizing a multitude of chemicals, olefins, and aromatics.

The three main activities of the petrochemical industry are production, transportation, and delivery of service to the end user. The competition within the energy service sector depends on the way companies can operate in each of these activities. Companies have to work within specified timeframes to get returns on their investments. Their infrastructures for production and distribution of their products are established based on these specified timeframes.

Developing marketing and distribution plans, pricing strategies and distribution strategies, and adopting business practices for strategic advantages and sustainable functioning all depend

The Role of Big Oil

on investment returns. Petrochemical organizations work within national and international regulatory frameworks. This is done to allow for information systems that are non-discriminatory and transparent and so that pricing policies, service standards, infrastructure specifications, and technical standards can be developed. It also encourages the adoption and development of fair business practices.

The continued growth of economies in countries such as China, India, and Brazil has increased demand and consumption for raw materials and fuels such as oil and gas.

The demand for gas in China is expected to double over the next five years alone, according to research by the International Energy Agency (IEA). China, whose economy continues to increase and proliferate, is set to become the world's third largest importer of gas, behind Europe and Asia Oceana. Meanwhile, overall global gas demand will increase by 2.7 percent per year.

Authors and research professors Susan A. Brown, Norman L. Chervany, and Bryan A. Reinicke discussed the dissemination of new information technologies and the marketable features and timeliness of innovation within IT organizations. They found that a key factor in the implementation of IT innovation must be support at the executive management level. Identifying problem areas such as organizational commitment, planning, infrastructure, and communication aided in defining challenges which were inherent, even within organizations that were viewed as IT innovation leaders within the industry.

Businesses in the English-speaking world are not ignoring the growing demand for oil and gas elsewhere. A recent Ernst & Young Capital Confidence Barometer report finds that 31 percent of oil and gas executives will be looking to make a purchase in the coming year. Many share the view that the overall global economy is improving and want to ensure they are well-positioned to take advantage of the growth in demand for the raw materials that allow emerging economies to expand their infrastructures.

Many organizations are also trying to settle on the business worth of information technology and its relationship between IT inputs and economic outcomes. Some studies imply that business executives and researchers have certain concerns about the value of IT innovation investments. The studies elaborate on how organizations can derive value from spending money on IT innovations.

When companies embark on oil and gas deals, one of the most common locations to do business is in the Middle East. The region is one of the richest when it comes to oil and gas reserves. Many of the largest oil companies in the world, measured by reserves, are based in Saudi Arabia (Saudi Aramco), Iran (National Iranian Oil Company), Qatar (Qatar General Petroleum Corporation), Iraq (Iraq National Oil Company), and Abu Dhabi (Abu Dhabi National Oil Company).

Working with a qualified language solutions provider, with expertise both in Arabic and in the oil and gas industry, is essential when embarking on deals and opening operations in the Middle East. Much of the material that will need to be translated for the oil and gas industry will be highly technical, whether it is an annual report; QHSE report; or the technical glossaries covering subjects like equipment, transportation and storage, flow metering, pipelines, exploration, drilling, and processing.

The unfortunate part of these global meetings and deals is that more jobs are being created outside the US as the outsourcing saga continues. This is increasing unemployment numbers here at home. (It makes one wonder, *What happened to the American dream?*)

As I discussed earlier, the petrochemical market is one of the complex markets in regards to production, processing, operations, and logistics for the delivery of goods and services to end users. The complexities are coupled with technology and regulatory compliances. In such scenarios, securing outcomes is becoming increasingly challenging and complex.

The petrochemical markets also have deep social and economic importance. These markets are complex and politically sensitive. The political sensitivity of the markets is especially high for economies dependent on other countries to maintain a supply of petrochemical products. And the prices of those petrochemical products have direct impact on the prices of products and services manufactured and delivered by those using petrochemical products.

To aid in all this, experts in providing complete translation and language solutions for global companies and law firms have been summoned. These professionals are proficient in serving the legal, financial, life science, software, and corporate markets. For example, Merrill Brink, with its thirty years of experience, offers a language solution, including translation, localization, desktop publishing, and globalization services.

The petrochemical industry has been recognized for its commitment to quality and its pioneering approach of leveraging technology to reduce costs, eliminate redundant processes, and speed up the translation life cycle. The petrochemical markets have been liberalized or are in the process of liberalization across the world.

In today's global society, vertically integrated monopolies have well-defined service territories and locked-in customer bases. However, this is giving way to more flexible market arrangements. The monopolies of the market have allowed petrochemical organizations to retain their power in different ways. They have the advantages of an established infrastructure, distribution network, and marketplace, as well as technological experience and knowledge, established networks, and understanding of the market. The competitive transformation of the petrochemical industry globally holds the potential to meet the demand for the petrochemical markets more efficiently and in a greener, more cost-effective manner.

The petrochemical industry also holds the potential for greater innovation in transporting, packaging, and delivering petrochemical products that can yield cleaner, more reliable, and more reasonably priced energy. In addition, policy and decision makers in global organizations continue to cite the need for changes within the business environment, which influences the needs in the information technology arena.

The success of IT innovation adoption models is focused on the commitments made by chief executive officers (CEO), chief information officers (CIO), managers, technology analysts, and leaders of IT organizations in a unified innovation adoption model. An existing general problem is that petrochemical companies operating internationally adopt and implement more IT innovation projects than do petrochemical companies operating primarily within the United States.

Bryant, Hull, and Enator conducted a pre-study analysis of the production data used within Chevron Angola, the purpose of which was to define the requirements for a replacement of the current production data accounting and reporting systems. My work in Angola was to evaluate the adoption of IT innovation and IT investment practices within the petrochemical industry, with a focus on internationally operated subsidiaries. The precursors in my study of organizational support, organizational climate, IT innovation, and IT investment practices were related to the expanded, working capital allocated to petrochemical companies operating outside the United States.

Since the successful implementation of IT technology and innovation requires support from leaders of the specific field and management, support was important for the study to evaluate the roles of different professionals in the success of technological innovations within organizations. The study resulted in increased production and expanded working capital allocated to petrochemical organizations operating outside of the United States.

Independent variables encompass management support, organizational support, and employee IT adaptation and IT investment, which may relate to the expanded work capital allocated to petrochemical companies. The petroleum industry operates in a complex business environment. The complexity of this business environment comes from regulatory frameworks; technology; and political, social, and environmental factors.

Chapter 5

BIG OIL'S INFLUENCE ON INFORMATION TECHNOLOGY

IN RECENT TIMES, oil has been dominating the world in more ways than we could ever think of. It is the bread and butter of the global economy and the catalyst for economic prosperity—the richer get rich and the poor ... (well, you know how that goes). Many would agree that without the petroleum industry, the technological era wouldn't be where it is now. IT's triumph and advancement in its fast-growing and evolving intelligence has always been backed up by the demands to make more business, conduct more research, and create better solutions to formulate cleaner oil and gas.

The organizations I studied have pursued cooperative agreements to obtain fast access to new technologies or markets, allowing them to profit from scale economies in joint research and production. These arrangements can be made with competitive partners or companies operating in different business environments.

Corporate agreements help organizations to gain sources of knowledge outside the organizations' boundaries. These organizations share risk together and contract with each other for complementary skills and work, which reduces the costs that each individual organization might have borne had they done research independently. Through these partnerships and alliances, companies have access to the resources they need, which helps companies to learn new capacities and ways to use IT. The transfer of technical knowledge and expertise helps to improve an organization's position to serve the market. With this strategy, organizations not only learn to manage their dependence on other organizations but also to maintain parity with their competitors.

IT innovations in organizations can be categorized into three distinct arenas: innovation that occurs within the IT function, innovation that occurs at the individual user or work group level, and innovation that happens at the organizational level. The emphasis within the literature is on IT innovation adoption practices.

Innovation diffusion theory implementation represents the infusion stage in the process of IT innovation diffusion. Once innovation adoption has occurred and has been applied to information technology, organizations can proceed to incorporate IT innovation within their daily operations and to support operational job tasks and IT innovation adoption practices.

The focus of an IT innovation program should be on achieving goals and objectives. Continuous improvement means the activity is recurring to increase the ability to fulfill requirements. At each level of the organization, continuous improvement takes place on an ongoing basis to involve the employees related to it and help manage day-to-day activities. People can participate and contribute to achieve innovation objectives through the employee involvement process.

The decentralization of foreign research and development allows for the tapping of different national systems for IT innovation and allows for various aspects of technological diversification that are referenced within the community market. Information technology innovation leads to a diverse stream of thoughts, processes, products, and technologies and creates additional expressions and efficient interactions, all while tumbling developmental expenses. It also can capitalize on site-exclusive compensation through the international distribution of employment among overseas research and development labs and improve reactions to area requirements regarding time and significance in a characteristic approach to area technological collaboration.

As knowledge increases in a fast-moving society, the benefits and costs of IT are rapidly coming into focus. IT brings many advantages to individuals and organizations, including helping with the costs and accessibility of information and other productivity benefits. For example, unprecedented amounts of data are now available in digital form, simply because information is easily and inexpensively manipulated, stored, and accessed digitally. We can access things as simple as magazine subscriptions or things more complex that we don't want others knowing, like employment histories and cash withdrawals.

The competition within the energy service sector depends on the ways companies can operate. Companies have to work within specified timeframes to get returns on their investments and when establishing their infrastructures for production and distribution of their products. Their development of marketing and distribution plans, pricing strategies, distribution strategies, and business practices for strategic advantages and sustainable functioning all depend on investment returns.

Petrochemical organizations work within national and international regulatory frameworks. This allows for information systems that are non-discriminatory and transparent in relation to pricing policies and so that service standards, infrastructure specifications, and technical standards can be developed. The adoption and development of fair business practices adhere to these regulations also.

The petrochemical market is a complex market as far as production, processing, operations, and logistics for the delivery goods and services to end-users. The complexities are coupled with technological and regulatory compliances. In such scenarios, securing outcomes is becoming increasingly challenging and complex.

The challenges of the petrochemical industry across the world are not just limited to meeting the demands of the marketplace. They also have to deal with the monopolistic nature of the industry. These types of challenges pose some of the greatest hindrances for IT innovation and performance improvements.

The petrochemical industry is both technology intensive and knowledge intensive. The importance of the role of technology to continue the operations as well as improve efficiency is high. But in order for the industry to continue to move forward, various factors, including competition, government controls and regulations, and national and international marketing forces, must work together to improve efficiency and contribute to innovation.

Chapter 6

LIBERALIZATION IN THE PETROLEUM INDUSTRY

LIBERALIZATION REFERS TO a relaxation of previous government restrictions or guidelines, usually in such areas of social and economic policies. Liberalization plays an important role in the improvement of the petroleum industry; it has changed the outlook of energy that services markets across the world. But the influence of liberalization is not limited to the market. It also affects the entire value chain of the industry.

> On the other hand, liberalization should provide a company that regains its innate entrepreneurship, and uses it fully to renovate its management, with a golden opportunity to expand the foundation of its business operations and strengthen its corporate structure.
> —Takeo Kikkawa
> Professor, Doctor of Economics

In several markets, liberalization introduced competition. This resulted in better utilization of strategic resources and a better quality of products and services. Ultimately, it has improved customer service and satisfaction. The un-liberalized market is rigid, monopolized, and not flexible enough to provide choices for suppliers and consumers. The liberalized market successfully addresses these issues. Customers now have the flexibility of choosing suppliers, products, quality and pricing, location options, and outlets that meet their needs.

The industry has started working toward improving efficiency and the optimum utilization of resources. The results are visible in all functional areas of petrochemical products, including production, transmission, distribution, services offered, marketing strategies, and efficient technologies in which different developmental and innovative activities are taking place.

In his speech to the International Forum on Globalization, Victor Menotti, the Executive Director of the International Forum on Globalization (IFG) since 2009, stated that energy experts believe this is the era of a historic transition phase for the energy industry. The liberalization process is full of challenges from technical, social, economic, and political perspectives. Different types of trade barriers hinder the process of liberalization. These barriers have been identified by national and international institutions worldwide, and estimating the cost factors of each of these barriers is still a challenge.

Any government policies or regulations restricting trade among nations are considered trade barriers. Trade barriers can be technical or come in the form of import duties, quotas and licenses, export licenses, tariff barriers, subsidies and incentives, nontariff barriers to trade, Voluntary Export Restraints, and others. The World Trade Organization (WTO) has addressed the issue of trade barriers in various articles of The General Agreement on Trade in Services (GATS). In Article XVI, Market Access of GATS WTO in compliance with Article 1, states, "Each WTO member shall accord services and service suppliers of any other member treatment no less favorable than that provided for in the Member's schedule. Where market access commitments are undertaken, Members shall not maintain such measures as, inter alia, quotas or other limitations on the number of service suppliers; limitations on the number of natural persons employed in a particular service sector; or limitations on foreign capital participation."

The United States is one of the largest producers and its people some of the largest consumers of energy products. The US has adopted various policies and regulatory reforms proactively, including the US National Energy Policy. This policy has a consistent theme of liberalization of the energy market and has facilitated competition through regulatory changes and reform. It was initiated within the Carter Program for oil, which allowed liberalization of the oil market and addressed the energy shortages. With the Natural Gas Policy Act of 1978, the US reduced the regulation on the gas market. Then the establishment of the Strategic Petroleum Reserve (SPR) "as a means of giving both state and private actors' greater control over mitigating short-term supply disruptions." This supported state and private owners so they could meet short-term supply shortages.

The National Energy Policy places an important emphasis on increasing the domestic production of energy and economic growth. The policy also provides a framework for global energy environmental policies, energy conservation, non-fossil fuel sources, local and regional pollution, transportation, research and development on clean energy technologies, and clean energy technologies awareness abroad.

The United States governmental policies focus on energy efficiency, environmental sustainability, development, and the promotion of renewable energy services. The country adopted regulatory action in 1985 and 1992, and undertook deregulation of gas markets during the same time period. The United States has also initiated programs to keep its supply and demands balanced. The Federal Power Act and the Energy Policy Act of 1992 were adopted and have been important developments in the Federal Energy Regulatory Commission (FERC), which is a regulatory body for wholesale and retailing of electricity services.

Even though many organizations view information technology as an economic development process, some researchers advocate innovations for creating rapid and effective communication aimed at reducing costs for IT innovations within the global business environment. Different approaches of innovation have been developed at various times. These approaches toward innovation reflect the transition of the manufacturing-based economy to a services-based economy.

As I mentioned before, a key factor for the implementation of IT innovation focuses on acquiring support at the executive management level. Identifying problem areas, such as organizational commitment, IT infrastructure planning, and communications, aids in defining the challenges inherent to IT changes.

Information technology innovation adoption scholars have recently expressed concern for the dominance theories, which has caused a high degree of enforcement, conformity, and lack of IT innovations. Innovation in information technologies has transformed organizational IT structures.

The methods vary in which the value of IT innovations investments can be assessed within organizations. Profit-seeking organizations continue to invest in IT innovations. But the results of empirical searches for IT value are often mixed and lead some to argue that IT innovation is a commodity input that does not matter.

Wright suggested that the success of any organization or firm depends on the development, integration, and exploitation capacity of real flows of knowledge within that organization. On the other hand, authors and researchers Wonseok Oh and Alain Pinsonneault noted that there are various articles and research studies that compared two conceptual and two analytical approaches for assessing the strategic value of IT. Alignment between business strategies and IT systems strategies as a cost reduction method was found to have a negative association with organizational expenses. And according to authors and researchers Jeffery Pfeffer and Richard Salancik, companies can stabilize in a turbulent environment through cooperative agreements.

Again, the petrochemical industry is part of the biggest businesses in the world. Liberalization opened various markets and increased the competition. These organizations have to work within specified timeframes to get the projected returns from their investments, to establish their infrastructure for production, and for the distribution of products. And national and international pressure for maintaining social responsibilities and adopting technologies that can provide sustainable utilization of resources and environmentally friendly procedures are increasing.

Companies in the petrochemical industry face both the demands of the energy marketplace and the monopolistic nature of the industry. This nature poses a hindrance in the way IT innovation and performance improvements are accomplished. In addition, sources indicate that the performance of petrochemical companies can be improved when certain changes happen related to how governments deal with companies. It helps when monopolies are dissolved and competition is introduced within the marketplace.

In the future, we will justify how much liberalization has changed the outlook of the energy services markets across the world. Yet the influence of liberalization is not limited to the petrochemical market; liberalization affects the entire value chain of the petrochemical industry.

Chapter 7

OIL PRODUCTION INFLUENCE ON IT OUTSIDE THE US

BOTH THE PETROLEUM and IT information industries hold keys for a better future. But since oil production in the US and other countries is expected to enter terminal decline, many researchers have conducted and are conducting data collection and observation to ensure further understanding of how to deal with the ongoing problems.

Sometime during 2008, oil prices reached a record high of $145/barrel. Alternative oil was introduced—particularly the use of ethanol—but it backlashed with unintended consequences of creating higher food prices, particularly in developing countries.

Optimists forecast the decline in oil production will begin after 2020. They also assume that major investments in alternatives will occur before a crisis hits the world economy. Thus, they feel the lifestyles of nations that are heavy oil consumers won't experience major changes. However, we cannot dismiss the fact that the wide use of fossil fuels has been one of the most important stimuli of economic boom and prosperity since the

Oil Production Influence on IT Outside the US

Industrial Revolution, allowing humans to participate in the consumption of energy at a greater rate than it is being replaced. Many believe that when oil production decreases, human culture and modern technological society will be forced to drastically change.

The information we have covered thus far has provided a sort of road map of US and global production declines. This should show us where production is headed. A logical conclusion tells us that oil is getting scarcer and prices are going up. As prices go up, so does inflation. So what's the best way to fight back inflation? There are number factors to consider when developing strategies to fight inflation. Countries should invest in commodities such as gold, oil and metals; historically, the prices of commodities rise greatly during inflation.

Oil investments are at an all-time high. There is an abundance of oil and many gas projects on the market today. Jobs are being shifted overseas to cater to mogul companies, whose intentions are to pay cheap labor and make double the profit.

Though many IT jobs are also being shifted overseas to help build businesses, US salaries are not as bad as in other sectors.

Finding oil is not piece-of-cake work, and it can be extremely technical. Companies that employ integrated exploration techniques in their program generally do so through the course of time, resulting in the companies coming out as winners. The successful exploration tools of the future will not be a single tool or technique. Success will be measured through the ability to properly integrate, develop, and implement logical exploration models based on high-quality data acquisitions.

The importance of knowledge increases with the growth of the industry. In today's scenario, the industry has become complex and bigger. A demand for expertise has increased, but the sources of expertise are dispersed. Learning networks have become more active than individual firms.

The dynamics of innovation have changed. Innovation is the result of internal knowledge development but also external knowledge acquisition and use. External sources of knowledge are important for the innovation of new products. This transformation of knowledge is based on the resource-based view of the organization.

The economic activities and populations of countries like India and China are growing. Governments face challenges in providing for the needs of their people so they can achieve the basic level of comfort. In the United States, corporations plan to outsource many thousands of IT jobs. Most of these jobs will belong to so-called offshore organizations in India or Southeast Asia. The media buzz and corporate momentum around IT offshoring and outsourcing continues and shows no signs of abating.

It is true that the nature of the work will change dramatically, and it may prove wholly undesirable, depending on one's viewpoint. IT salaries will increase in countries like Asia and India and will decrease in the US. Scary? Yes, but there is no need to panic.

A series of research projects have been conducted to provide answers to the growing concerns of IT jobs threatening to diminish in the US and how it would impact our lives. Chevron is one of the largest oil producers in the world and has conducted its own research. Respondents from various geographic locations within the company were surveyed.

The results of the IT effectiveness survey of ten petrochemical employees provided empirical results. The results of this study indicated the correlation between management support and IT investments is significant, suggesting a relationship exists between management support and IT innovation investment within the petrochemical industry. As management support increases, IT innovation investment increases among organizations within the petrochemical industry.

Chevron processes petrochemical products, among many other ventures, and is a recognized leader of innovation. Chevron's motivation and sense for global business serves as the foundation for the future of the company. It stands behind the quality of service the company provides. Chevron's mission is "to be the global energy company most admired for its people, partnership, and performance" and to create an organization that learns faster and better than its competitors. Chevron plans to become the top petrochemical company by establishing global benchmarks; sharing and implementing best practices for organizational goals and objectives; learning from experience; and continuing to learn through customer needs, technology, and personnel. The company also believes in creating partnerships based on integrity and high performance, and its success relies on aggressiveness, self-management, and constant self-improvement. The company's vision reflects its desire to deliver a level of quality surpassed by no other in the industry. These values solidify Chevron on a global scale.

Social Interconnection

Social interconnection is one of the major determinants associated with the rate a technological innovation goal is adopted. Social closeness helps in integrating goals to a common goal. It also affects the innovativeness of individuals. "Learning transformation is not just the collection of job techniques and information but also a procedure of becoming a specific individual or equally, to avoid becoming a specific individual, it is founded in the forming of an identity that implies learning can become a source of meaningfulness and a personal or social energy."

—Étienne Charles Wenger,
Educational theorist and practitioner

Chapter 8

INFORMATION TECHNOLOGY VALUE

I ONCE HAD a conversation with a high-ranking member of the business unit for the company I work for. He was questioning the value of IT to the entire organization. He viewed IT as an insignificant entity of the organization.

I asked him a very simple question: "What would you do if you woke up one morning and the entire IT department was gone? Could the organization survive?"

Before he could provide a reply, I stated that he would no longer receive email (which in his case and mine might be a good thing) and would not have access to all the charts and graphs that are stored on network resources, meaning he could not perform his job.

So the question is: "Does IT really matter?"

IT innovation leads to a diverse stream of thoughts, processes, products, and technologies and creates additional expressions and efficient interactions. It also tumbles developmental expenses. It can capitalize on exclusive compensation through the distribution of employment among international research and development labs and can improve reactions to area requirements in provisions of time, significance, and characteristic approaches to area technological collaboration. However, the economic cycle has implied considerable impact on innovation initiatives. The economic downturn and recession encouraged IT cost-cutting efforts that may direct organizations toward withdrawing from or not investing in research and development of new technology.

In the nineteenth century, the British Industrial Revolution was founded on the job apprenticeship system, during which a skilled workforce transferred its knowledge of its craft to the next peer group. Years later, in the US, the IT innovative revolt was characterized by the elevation of specialized executives directed by the disconnection involving tenure, domination, and the evolution of graduate level, executive learning. Toward the conclusion of World War II, organizations industrialized by the US encompassed authority supervisory corporations whose primary objective was the advancement of modern technology. Japanese corporations later challenged the US's industrialized organizations in the same industrialized sectors in which American organizations had previously acquired an enormous competitive advantage.

The most important role of IT innovation in historical, empirical methodology is to function as a benchmarking instrument for a pragmatic examination of the innovation processes within a current, national, socio-economic scheme. IT has been considered a sophisticated and competitive tool for gaining a strategic advantage within the present business environment. Virtually no area of business management in which IT is not a key component either directly or indirectly exists. Information technology allows organizations to contend with daily activities for the more specific and highly complex activities of their particular businesses.

IT tools are directly related to the organization and industry that is conducting business. The more an organization is in line with the predefined IT abilities, the closer that organization will be to achieving its financial goals. IT concepts should be associated with innovation so that investments in innovation activities can be optimized.

The methods in which the value of IT investment can be assessed within an organization vary. As was mentioned previously, profit-seeking organizations continue to invest heavily in IT innovations in order to create organizational stability. Once innovation adoptions have occurred and have been applied to information technology, organizations can proceed to incorporate IT innovation within their daily operations and to support operational job tasks and IT innovation adoption practices.

Information technology has provided an effective platform and means for people around the world to learn and develop skills and knowledge. However, to use this platform, one needs to understand it first. This requires a facilitator or instructor who can teach and facilitate the basic learning process. Because information technology requires continuous improvement and training, the focus of the IT innovation program should be achievements of goals and objectives; continuous improvement means the activity is recurring to increase the ability to fulfill requirements.

Big Oil & the IT Industry

Chapter 9
OPERATION BIG BOYS

THE UNITED STATES dominated world oil production in the first half of the twentieth century. US fields accounted for slightly more than 70 percent of the world oil production in 1925, 63 percent in 1941, and over 50 percent in 1950. The US oil industry operated in a unique regulatory environment that included a permissive legal regime, generous tax treatment, and a cooperative system of national production control that was centered on the state of Texas, which accounted for almost half of total US production.

The strong position of the US in world oil provided multiple advantages. In addition to the US being central to military power and economic prosperity, oil control gave the US leverage over its allies and its former and prospective enemies. In order to maintain this leverage, US policymakers saw growth as essential to preventing the recurrence of the divisive, ideological, and social conflicts of the interwar years.

To fuel economic recovery and to prevent Western Europe from becoming dependent on the Soviet Union for energy, the US sought to ensure that this critical area received the oil it needed. In turn, economic growth was crucial to mitigating the divisive class conflicts that had divided European and Japanese societies in the first half of the century. By controlling access to essential oil supplies, the US was able to reconcile its goal of German and Japanese economic recovery and integration into a Western alliance that ensured against the recurrence of German and Japanese aggression.

Economic growth in Western Europe and Japan was also central to the containment of Soviet power and influence during the cold war because it helped prevent these areas from falling to communism through internal processes.

After World War II, the Soviet Union lacked sufficient oil to fight a major war. They were hit hard by wartime damage, disruption, transportation problems, and equipment shortages. Overseas, Soviet oil production dropped after the war. The Soviet Union was a net importer of oil until 1954.

Exclusion of the Soviet oil production from the Middle East retained oil for the Western recovery and kept the Soviets short of oil. US and British strategic planners wanted to keep the Soviets from the Middle East because the region contained the most defensible locations for launching a strategic air offensive against the Soviet Union in the event of a global war.

In truth, oil production is a mixed bag. There is no question that natural gas production on private lands increased by 16.4 billion cubic feet per day from fiscal year 2005 to fiscal year 2011. Meanwhile, natural gas production on federal and Indian lands has steadily fallen. This trend began around the fall of 2002 and was due to a consistent decrease in offshore gas drilling. However, such gas production onshore, on Federal lands, is actually higher now than it was at the end of the Bush administration.

Moving IT Forward

To review, the focus of IT innovation programs should be on achieving goals and objectives. Continuous improvement means the activity is recurring to increase the ability to fulfill requirements. The commitment and personal involvement of top management helps in creation and deployment of clear goals for IT adoption processes. Top management needs to be consistent with the objectives of the petrochemical organization; this inconsistency is one of the main reasons why organizations fail to thrive in competition.

A dedicated organization management team must be committed to excellence. Organizational executives must explicate obvious initiatives and encourage an organizational society to accomplish these goals."

—Y.K. Shetty
Professor of Management,
Utah State University's College of Business

There is no secret, witch potion, or magic spell that will make a business a successful one. But well-developed strategies, planned goals, and employees empowered with IT are part of the success of any organization.

Employee empowerments interventions are defined as the process of giving employees the power to make decisions about their work environments. Employee empowerments interventions are keys to the success of any IT adoption process. Empowerment allied with a quality improvement agenda may also provide an opportunity for employees to outline their work environment to improve quality and customer satisfaction. And a democratic management style allows employees to pursue their personal aspirations within a given context and to contribute to the efficiency of an organization. The level of participation in a democratic management style is higher than in any other management style, even though scholars have criticized the democratic management style.

It is important to note that a strong orientation for individual progression and growth is observable in occupational organizations, particularly organizations motivated by humanistic, democratic, or spiritual ideologies. Members of such organizations may be encouraged to follow their individual aspirations to such a degree that the potential of the organization to escalate efficient, collective achievement is compromised, eventually threatening the organization's survival.

Since stakeholders in a company include all customers, employees, investors, suppliers, vendors, and other related people, their satisfaction is the central theme of the IT adoption process. Ultimately, all activities of an organization are directed toward profitably satisfying consumers. Stakeholders' satisfaction shows in reductions in errors and complaints, increasing the quality of products and services, and offering customers better products and services to meet or exceed their expectations. Customer feedback should be collected regularly and adopted in the product development.

Culture Change

Petrochemical organizations usually have various divisions under the corporate banner. Culture change requires excellence in all aspects of an organization's operations, with services being administered correctly in the beginning and defects being eradicated from daily job activities. Cultural transformation is achievable, and authentic intent to revolutionize processing systems is a significant aspect of success when adapting IT innovations.

The concept of competition has changed within the last few years. Research and development (R&D) cooperation strategies have been highlighted. These R&D partnerships are based on various corporate objectives, and certain connections between organizations have been prevalent within the last few years. There are numerous reasons supporting this type of inter-company collaboration. Some examples of collaborations are: the combined efforts toward research and controlled diffusion of expertise, the formation of an innovative market, and discovering a market segment.

Another reason for contributions to knowledge sharing by organizations is strategic collaboration in the product management process—like the knowledge and competences transfer as mentioned in the work of Bernard L. Simonin, Professor of Marketing and International Business at Tufts University.

Knowledge Is King

Competitors are investing in various research and development programs and adopting new strategies to maintain or gain market shares. When customers are more informed about the options available to them, they make smarter choices.

One major source of knowledge for competitors is the employees of another company. The transfer of skilled workers or employees leads to a loss of trained employees and sometimes knowledge and information going to other companies. Organizations are developing strategic alliances to have complementary technology and to reduce the use of the innovation cycle. Companies can share knowledge and expertise, along with shortening investments and cycle time.

The risk of failure and uncertainty are also shared. As an organization increases its knowledge and expertise, the innovation cycle duration decreases. Efforts have been directed toward finding new processes that are smaller and more cost effective. The new processes have resulted in reductions of the innovation cycle.

In the present business environment, knowledge has been perceived as an important asset and the most important resource of organizations. In cases of technologically complex businesses, the role played by knowledge becomes crucial.

There are two kinds of knowledge within an organization. First is tacit knowledge. This is knowledge received from academicians and researchers for the ability to maintain the competitive advantage. Oxford University Press authors Ikujiro Nonaka and Hirotaka Takeuchi have linked tacit knowledge with the second kind of knowledge: organizational learning and innovation. This knowledge is considered to be the greatest strategic resource of a company. Tacit knowledge is useful in maintaining the competitive advantage. Organizational learning and innovation are easily available, easy to share, easy to store, and best used to gain a competitive advantage.

Conclusion

Information technology is considered a sophisticated and competitive tool for gaining a strategic advantage within the present business environment. Numerous applications are available for various kinds of operations and objectives, stemming from simple call management systems within call centers to enterprise resource planning systems developed for large manufacturing organizations. Virtually no area of business management exists in which information technology is not a key component either directly or indirectly. Information technology allows organizations the ability to contend with daily activities and the more specific and highly complex activity levels of their particular businesses.

Some members of the IT community see IT innovation as part of a process of economic development and goals, and some members advocate IT innovation for creating rapid and effective communication directed at cost reduction. While the discovery of IT innovation becomes more valuable and helpful for many businesses, the petroleum industry has their eyes set on how to incorporate IT into the Petrochemical Arena. And the IT innovation environment possesses characteristics that may make the Petrochemical Arena prone to various IT innovation investments.

As the next generation of the digital oil field continues to merge with the IT system, the demand for bridges connecting oil and gas production systems with IT systems is becoming more evident. We must continue to find ways to integrate applications and devices into the oil and gas production environment. The demand to incorporate devices such as the iPhone, iPad, tablets, Blackberry, and Android devices is increasing across the oil and gas arena.

The future of petrochemical companies and IT systems moving forward into the twenty-second century will heavily

depend on the foundation that is being developed now. The new oil fields being presented are referred to as "ifield" and will offer numerous opportunities that will be immersed in transformative technology innovations developed as part of efforts to create the digital oil field of the future.

> We used technology to change what we do, rather than optimized what we have always done.
> —Jim Williams, Chevron manager

> My role going forward will be to continue to act as an agent for promoting the integration of IT systems with Oil and Gas Production systems. Where else can you go and get paid to play with million-dollar toys?
> —Dr. Byron K. Wallace
> Chevron IT Infrastructure Analyst

"Excellent, Easy Reading, and Good Humor . . ."
—Jim Pinto, Author of *Automation Unplugged: Pinto's Perspectives, Pointers, & Prognostications & Pinto's Points: How to Win in the Automation Business*

BIG OIL AND THE IT INDUSTRY

Two Giants Shake Hands

Dr. Byron K. Wallace

www.ingramcontent.com/pod-product-compliance
Lightning Source LLC
Chambersburg PA
CBHW040831180526
45159CB00001B/146